Science

New York City Edition

Reading Support and Homework

Grade 2

Harcourt
SCHOOL PUBLISHERS

Visit *The Learning Site!*
www.harcourtschool.com

Printed in the United States of America

ISBN 13: 978-0-15-381391-7
ISBN 10: 0-15-381391-1

1 2 3 4 5 6 7 8 9 10 1413 17 16 15 14 13 12 11 10 09

Contents

Chapter 3 Motion

Chapter 4 Living and Nonliving Things

Chapter 5 Plants

Chapter 6 Weather

VOCABULARY

Name ______________________________

Date ______________________________

Welcome to *Science!*

Find these things in your book.

1. What animal is on the cover of your book?

____

2. How many chapters are in your book?

____

3. Look on page 4. What word is in a yellow box?

____

4. How many questions are on page 9?

____

5. Look on page 19. What Inquiry Skill is shown?

____

6. Look in the Health Handbook. It is in the back of your book. On what page do you find a skeleton?

____

Name ______________________________

7. What is the first word in the glossary?

____

8. Look on page R26. Tell one science safety rule.

____

____

9. Find a picture of something you like.
Tell about it.

____

____

Name ______________________________

Date ______________________________

Writing Practice

Tell a Recycling Story

Write to Express–Narrative

Suppose you are something that will be reused in a recycling project. Write a story that tells what you were first used for, how you were recycled, and what you are used for now. Use the story map below to help you plan your story.

What are you? ______________________________ ______________________________
What were you used for first? ______________________________ ______________________________
How were you recycled? ______________________________ ______________________________
What are you used for now? ______________________________ ______________________________

Name ______________________________

Date ______________________________

Vocabulary Power

Exploring Earth's Surface

Exploring is another way to say finding out about. Earth is another way to say our planet. So exploring Earth is another way to say finding out about our planet.

Match the words to the phrases.

erosion•	• dirt
earthquake•	• shaking ground
boulder•	• wearing away
soil•	• big rock

Rewrite the sentences with different words.

1. A lot of dirt was lost due to wearing away of the land.

____

2. The shaking ground moved the big rock.

____

Name ______________________________

Date ______________________________

Lesson 1 - What Changes Earth's Surface?

1. **Inquiry Skill Practice–Observe**

The pictures show how a rock changed over time. Observe them carefully. Then write a caption for each picture.

______________ ______________ ______________

______________ ______________ ______________

______________ ______________ ______________

2. **Use Vocabulary**

Match the term to the words that tell about it.

erosion •	• wind and water break down rock into smaller pieces
weathering •	• a place where lava comes out of the ground
earthquake •	• wind or moving water moves sand and small rocks
volcano •	• a shaking of Earth's surface

Name ______________________________

3. Focus Skill **Cause and Effect**

Use this chart with the Reading Review.

cause		effect
rain and Ⓐ ______________	→	weathering
sand or rocks moved by water or wind	→	Ⓑ ______________
an earthquake	→	a rise or fall of Earth's Ⓒ ______________
a volcano	→	a Ⓓ ______________

4. **Critical Thinking and Problem Solving**

How can waves at a beach cause erosion?

____

____

____

Name ______________________________

Date ______________________________

Lesson 2 - What Are Rocks, Sand, and Soil?

1. Inquiry Skill Practice–Draw Conclusions

Each picture shows a tree. The caption tells about soil where the tree grows.

Black Spruce Tree

wet soil

Red Oak Tree

rocky soil

Sand Pine Tree

sandy soil

What conclusions can you draw from the pictures and captions?

____

2. Use Vocabulary

Match the term to the words that tell about it.

mineral •	• bits of rock mixed with matter that was once living
boulder •	• solid matter that makes up rocks
soil •	• a very large rock

Name ______________________________

3. Focus Skill **Compare and Contrast**

Use this chart with the Reading Review.

Rocks

alike	different
They are pieces of (A) ____________ crust.	They can be different colors.
They are made up of (B) ____________.	They can be large or (C) ____________.
	They can be rough or (D) ____________.

4. **Critical Thinking and Problem Solving**

What can you tell about a plant that grows in soil that holds little water?

____

____

Name ______________________________

Date ______________________________

Lesson Quick Study

Lesson 3 - What Can We Learn from Fossils?

1. Inquiry Skill Practice–Communicate

Suppose you are a scientist. You uncover three fossils. Communicate what each one is and whether it is from a plant or an animal.

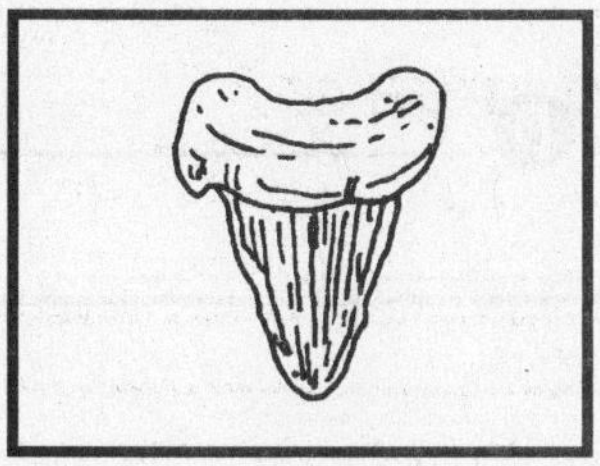

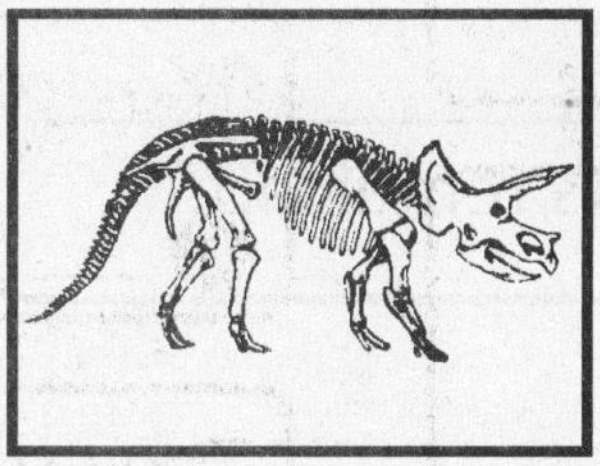

____________ ____________ ____________

2. Use Vocabulary

Write the term that best completes each sentence.

dinosaur	**fossil**	**extinct**

A ____________ is what is left of an animal or a plant that lived long ago.

A ____________ is an animal that lived on Earth millions of years ago.

Animals that died out are ____________.

Name ________________________________

3. Focus Skill **Sequence**

Use this chart with the Reading Review.

How a Trilobite Fossil Forms

First, a trilobite Ⓐ ______________. Mud and sand covered it.	→ Next, its Ⓑ ______________ parts rotted away and its Ⓒ ______________ parts remained.	→ Last, the mud, the sand, and the hard parts of the trilobite slowly turned to Ⓓ ______________.

4. **Critical Thinking and Problem Solving**

What can you tell from this fossil about where the animal lived and how it moved? How could you learn more about the extinct animal?

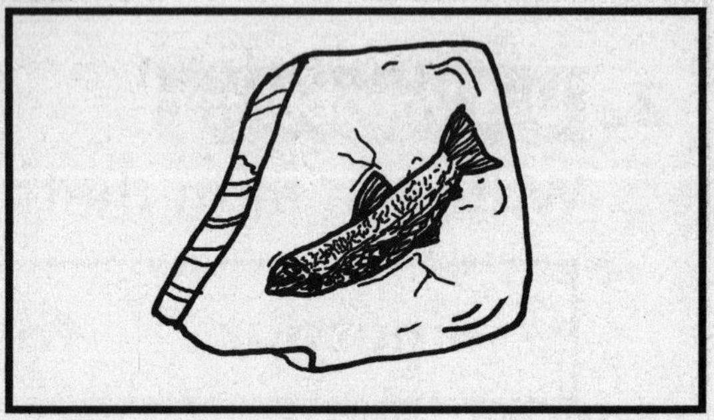

____

____

____

____

Name ______________________________

Date ______________________________

Natural Resources

Look at the words below. All of them have to do with natural resources. But some are words for nouns. Others are words for actions.

nouns		**actions**	
animals	trash	save	reduce
water	air	protect	help

I do not waste water. I protect natural resources.

Draw a picture to go with each word below. Then circle the nouns. Underline the actions.

pollution	**reuse**
recycle	**resource**

Name ______________________________

Date ______________________________

Lesson Quick Study

Lesson 1 - How Can People Use Natural Resources?

1. Inquiry Skill Practice–Draw Conclusions

Draw conclusions about the way this family uses plants.

____

____

2. Use Vocabulary

Write the term that best completes each sentence.

natural resource	resource

Anything that people use to meet their needs is a ____________.

Trees are a ____________ because they come from nature and are used in many ways.

Name ______________________________

3. Focus Skill **Main Idea and Details**

Use this chart with the Reading Review.

Main Idea and Details

People use natural resources to meet their needs.				
People breathe Ⓐ ______.	People use Ⓑ ______ to clean.	People use Ⓒ ______ to get metal.	People use Ⓓ ______ to make bricks to build with.	People use Ⓔ ______ to make clothes and paper.

4. **Critical Thinking and Problem Solving**

How are these things alike?

____

____

____

Name ______________________________

Date ______________________________

Lesson Quick Study

Lesson 2 - How Can People Harm Natural Resources?

1. Inquiry Skill Practice–Observe

Observe this park. Mark an X on all the things that people can clean up.

2. Use Vocabulary

Match the word to the picture that shows its meaning.

pollution •

• 

• 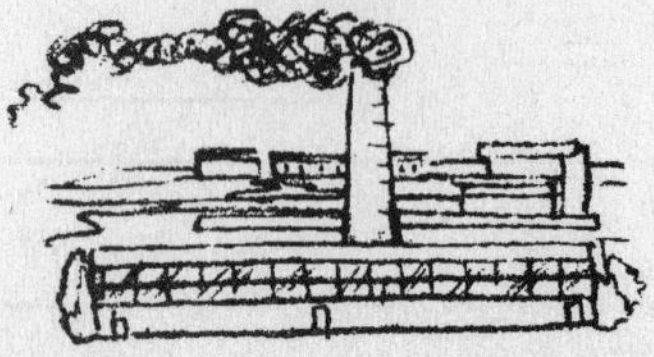

Name ______________________________

3. Focus Skill **Cause and Effect**

Use this chart with the Reading Review.

cause	effect
smoke and fumes	A ______________ pollution
factory waste and oil spills	B ______________ pollution
trash not in trash cans	C ______________ on the ground
too many trees cut down	D ______________ of natural resources

4. **Critical Thinking and Problem Solving**

Suppose you wanted to build a house on this land and still protect its natural resources. What could you do? How would it protect the natural resources?

____

____

____

Name ______________________

Date ______________________

Lesson Quick Study

Lesson 3 - How Can People Protect Natural Resources?

1. Inquiry Skill Practice–Plan an Investigation

Write the steps you would take to make a bird feeder. Draw a picture to show how your finished bird feeder would look.

____

____

____

2. Use Vocabulary

Write the word that best completes each sentence.

endangered	**reuse**	**reduce**	**recycle**

You can ____________ a food jar by storing buttons or marbles in it.

You can ____________ things made of glass and paper so that they can be used to make new things.

People help ____________ air pollution by riding a bike.

People can help ____________ animals.

Name ________________________________

3. Focus Skill **Cause and Effect**

Use this chart with the Reading Review.

cause	effect
People reuse things.	They save **A** ____________. They make less **B** ____________.
People recycle things.	They use old things to make **C** ____________ things. They make less **D** ____________.

4. **Critical Thinking and Problem Solving**

Suppose you are on your town council. What laws would you pass to protect the natural resources of your community?

____

____

____

____

Name ______________________

Date ______________________

Writing Practice

Tell How to Ride a Scooter

Write to Inform–How-To

Write a paragraph that tells a younger child how to ride a scooter. Finish the sentences in the outline to help you think of three steps. Use motion words from the word bank for Steps 2 and 3.

push	**pull**	**roll**	**turn**

Step 1
Start with one foot . . .
Step 2
Move forward by . . .
Step 3
To change direction, . . .

Name ______________________

Date ______________________

Vocabulary Power

Motion

Look at these words. How do they change when you add word endings?

magnet + –ic = magnetic **acid + –ic = acidic**	**observe + –ant = observant** **repel + –ant = repellant**

Now look at the words below. Add the new ending to each one to make a new word. Then use what you know to give the new word's meaning.

1. force + –ful = ____________

Circle the letter next to the new word's meaning.

A with no force

B with a lot of force

2. motion + –less = ____________

Circle the letter next to the new word's meaning.

A moving a lot

B not moving

3. speed + –y = ____________

Circle the letter next to the new word's meaning.

A with no speed

B with a lot of speed

Name ______________________________

Date ______________________________

Lesson 1 - What Are Ways Things Move?

1. Inquiry Skill Practice–Plan an Investigation

Look at these pictures. Use the items to plan an investigation to find out which ways the toy car can move.

Step 1: ______________________

Step 2: ______________________

Step 3: ______________________

2. Use Vocabulary

Write the word that best completes each sentence.

motion	**speed**

When something moves, it is in ____________.

____________ is how fast something moves.

Name ______________________________

3. Focus Skill **Main Idea and Details**

Use this chart with the Reading Review.

Main Idea and Details

Objects can move in different paths and at different speeds.

A toy car can move in a straight line.	A swing moves (A) __________.	The hands of a clock move in a (B) __________.	Objects can move quickly or (C) __________.

4. **Critical Thinking and Problem Solving**

If you are watching two people walk, how can you tell which one is walking faster?

____

____

____

Name ___________________________

Date ___________________________

Lesson Quick Study

Lesson 2 - What Makes Things Move?

1. Inquiry Skill Practice–Classify

Look at the pictures. Classify the movements as pushes or pulls. Mark an **X** in the correct box.

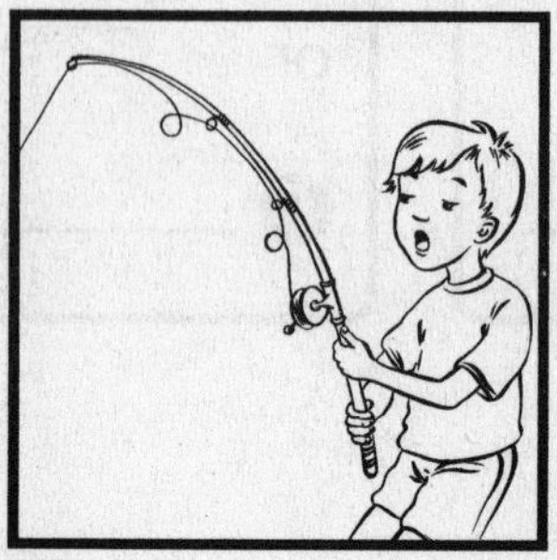

movement	push	pull
bringing in a fish		
hitting a baseball		
opening a door		

2. Use Vocabulary

Match the word to the words that tell about it.

force •　　　• a force that pulls things toward the center of Earth

gravity •　　　• a force that slows down or stops moving things

friction •　　　• a push or pull that causes something to move

Name ______________________________

3. Focus Skill **Cause and Effect**

Use this chart with the Reading Review.

cause		effect
An object is pushed or (A) ______________.	→	The object moves.
(B) ______________ pulls an object.	→	The object moves toward Earth.
Two objects rub against each other and cause (C) ______________.	→	Moving objects slow down or stop.

4. **Critical Thinking and Problem Solving**

Why is it easier for the first child to skate faster?

____

____

____

Name ______________________________

Date ______________________________

Lesson Quick Study

Lesson 3 - How Do Magnets Move Things?

1. Inquiry Skill Practice–Hypothesize

Look at the pictures. Hypothesize which object the magnet will attract. Explain why.

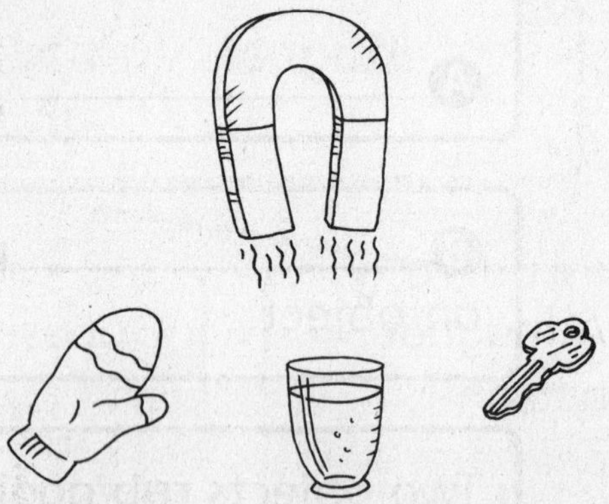

____

____

2. Use Vocabulary

Write the word that best completes each sentence.

magnet
poles
attract
repel

The two **N** poles on different magnets __________ each other.

All magnets have two __________.

Opposite poles on a magnet __________ each other.

A __________ can push or pull things made of iron or steel.

Name ______________________________

3. Focus Skill **Main Idea and Details**

Use this chart with the Reading Review.

Main Idea and Details

Magnets can pull iron or steel. Magnets can push and pull other magnets.

- A magnet has an N pole and an (A) __________ pole.
- Opposite poles (B) __________ each other.
- The same poles on different magnets (C) __________ each other.
- A magnet's force can move through air, water, and some (D) __________.

4. Critical Thinking and Problem Solving

Suppose you have two toy cars. You can pull the first car with a magnet but not the second one. What can you infer about the two cars?

____

____

____

Name ________________________________

Date ________________________________

Writing Practice

Describe a Plant

Write to Inform–Describe

Look at a plant. Then write sentences that describe the parts of the plant. Tell what each part looks like and what it does. Use the outline to plan your sentences.

Plant part	What it looks like	What it does

Name ______________________________

Date ______________________________

Vocabulary Power

Living and Nonliving Things

You can use these words together in sentences.

nonliving	**survive**	**shelter**	**nutrients**	**living**

Plants may need nutrients to survive.
A shelter may be made up of nonliving things.

Write a sentence that includes each word pair below. Use what you know about the words' meanings to help you.

1. living, shelter

____

2. living, survive

____

3. living, nonliving

____

4. nonliving, nutrients

____

Name ______________________________

Date ______________________________

Lesson Quick Study

Lesson 1 - What Are Living and Nonliving Things?

1. **Inquiry Skill Practice—Classify**

One picture shows a living thing. One picture shows a nonliving thing. Label each picture. Tell how you know what kind of thing each is.

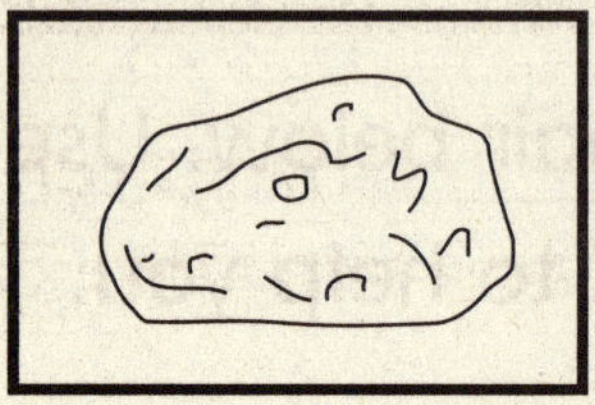

____________ ____________

____

____

2. **Use Vocabulary**

Write the word that best completes each sentence.

living
nonliving

A ____________ thing does not need food, water, or gases.

A ____________ thing grows and changes. It needs food, water, and gases to live.

Name ______________________________

3. Compare and Contrast

Use this chart with the Reading Review.

Living things	Nonliving things
A Living things ____________, food and water.	**B** Nonliving things do not ____________ food and water.
C Living things need ____________.	**D** Nonliving things do not need ____________.
E Living things make new ________________________ ________________________.	**F** ____________ things can not make new things like themselves.

4. Critical Thinking and Problem Solving

A deer and a flower are both living things. They both need water to live. How do they get water in different ways?

____

____

____

Name ______________________________

Date ______________________________

Lesson Quick Study

Lesson 2 - What Do Animals Need?

1. Inquiry Skill Practice–Observe

Observe the birds in this picture. Then fill in the chart to tell how some birds get what they need to survive.

How Some Birds Meet Their Needs	
food	
water	
shelter	

2. Use Vocabulary

Match the word to its meaning.

survive •	• a safe place to live
shelter •	• stay alive

Name ______________________________

3. **Main Idea and Details**

Use this chart with the Reading Review.

Main Idea and Details

Animals need many things to survive.

- Animals need space to move around.
- Animals need Ⓐ ________ to eat.
- Animals need Ⓑ ________ to drink.
- Animals need Ⓒ ________ to stay safe.
- Animals need Ⓓ ________ from air or water.

4. **Critical Thinking and Problem Solving**

How do a bear and a fish get oxygen, shelter, and food in different ways?

____

____

____

Name ______________________________

Date ______________________________

Lesson
Quick Study

Lesson 3 - What Do Plants Need?

1. Inquiry Skill Practice–Predict

Read about each plant, and predict what will happen to it.

Plant 1

Plant 2

Plant 1 gets enough water. It is in a sunny spot in the garden.

Prediction ______________________________

____

____

Plant 2 does not get enough water. It does not get enough sun in the garden.

Prediction ______________________________

____

____

2. Use Vocabulary

Match the word to its meaning.

nutrients •

• different amounts of water

• minerals

Name ______________________________

3. Focus Skill **Main Idea and Details**

Use this chart with the Reading Review.

Main Idea and Details

Plants need many things to grow and stay healthy.

Plants need water, Ⓐ __________, and __________ to make food.

Plants need Ⓑ __________ from the soil.

Plants need Ⓒ __________ to grow bigger.

4. **Critical Thinking and Problem Solving**

Suppose you had a plant that had gotten too big for its pot. What could you do to help it grow and stay healthy?

____

____

Name ______________________________

Date ______________________________

Vocabulary Power

Plants

When two words rhyme, they have the same ending sounds, but different beginning sounds. For example:

Pl<u>ant</u> rhymes with *sl<u>ant</u>*, *p<u>ant</u>*, and *c<u>an't</u>*.
Tr<u>ee</u> rhymes with *fr<u>ee</u>*, *m<u>e</u>*, *s<u>ee</u>*, and *b<u>e</u>*.

Look at each bold word. Write words that rhyme. Underline the letters in each word that stand for the ending sound.

roots	**flower**
____________________	____________________
____________________	____________________
shrub	**trunk**
____________________	____________________
____________________	____________________

Write a sentence that uses at least three words from the chart. Include one word that has to do with plants.

____

____

Name ______________________________

Date ______________________________

Lesson 1 - What Are the Parts of a Plant?

1. Inquiry Skill Practice–Observe

Plant A got sunlight, air, water, and nutrients. Plant B did not. Observe both plants and tell about the parts of each plant.

Plant A

Plant B

Observations: ______________________________

____

____

Observations: ______________________________

____

____

2. Use Vocabulary

Match the term to the part of the plant it describes.

roots •

stem •

leaves •

flower •

Name ______________________________

3. Focus Skill **Main Idea and Details**

Use this chart with the Reading Review.

Main Idea and Details

Each part of a plant helps the plant get the things it needs to live.

Roots take in **A** ________ and nutrients from soil.	A **B** ________ carries water to the leaves.	Leaves make **C** ________ for the plant.	**D** ________ help plants make new plants.

4. **Critical Thinking and Problem Solving**

Observe the plant. What might happen to it? Why?

____

____

____

____

Name ______________________________

Date ______________________________

Lesson 2 - How Do Plants Differ?

1. Inquiry Skill Practice–Classify

Use the chart to classify the leaves into three groups. List the trees the leaves come from in each group.

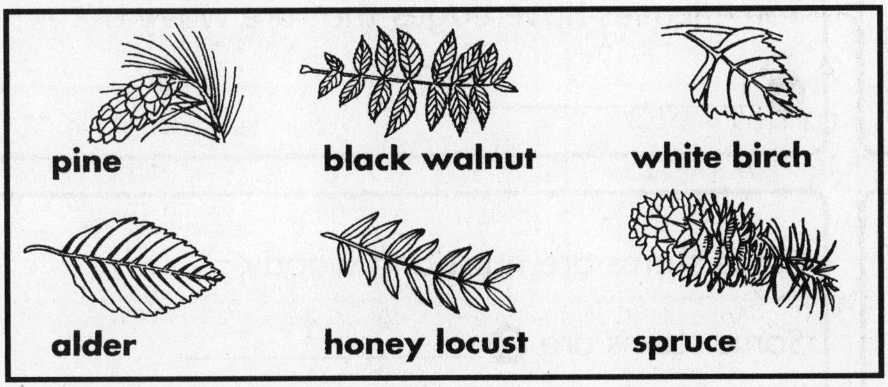

One Large Leaf	Many Small Leaflets	Needles

2. Use Vocabulary

Match the term to the sentences that tell about it.

shrub •	• It is hard and woody. It is the main stem of a tree.
trunk •	• It has many woody stems. It is also called a bush.

Name ______________________________

3. Focus Skill **Compare and Contrast**

Use this chart with the Reading Review.

Plants

alike	different
Plants have leaves.	Some leaves have smooth edges. Some leaves have edges that are wavy or (A) ____________.
Plants have stems.	Some stems are hard and woody. Some stems are (B) ____________.
Plants have roots.	Some plants have one (C) ____________ root. Some plants have many (D) ____________ roots.

4. **Critical Thinking and Problem Solving**

Suppose you and a friend want to find out whether two plants are the same kind of plant. What could you do to find out?

____

____

Name ______________________

Date ______________________

Lesson Quick Study

Lesson 3 - What Are Some Plant Life Cycles?

1. Inquiry Skill Practice–Communicate

Read about how a plant is growing. Then draw pictures to communicate what is happening.

________ ________ ________

________ ________ ________

________ ________ ________

2. Use Vocabulary

Write the term that best completes each sentence.

life cycle	**germinate**

Seeds need oxygen, water, and warmth to ________.

All the stages of a plant's life make up its ________.

Name ______________________________

3. Focus Skill **Sequence**

Use this chart with the Reading Review.

Life Cycle of a Bean Plant

A bean seed germinates. First, the Ⓐ __________ grows down. → Next, the Ⓑ __________ breaks the ground. → Then, more leaves and stems grow. The plant also makes Ⓒ __________ and seeds. → Last, a new Ⓓ __________ begins.

4. **Critical Thinking and Problem Solving**

What would happen if maple trees stopped making seeds?

____

____

____

Name ______________________________

Date ______________________________

Vocabulary Power

Weather

Context clues can help you understand the meaning of a new word. In the sentences below, time of year helps you understand what a season is:

Spring is Kelly's favorite season. She says it is a beautiful time of year.

Find context clues to figure out the meaning of each underlined word. Draw a line under the clues. Circle the letter next to the words that best complete the sentence.

1. It did not take long for the puddle to evaporate. It seemed to disappear under the hot sun. To evaporate is ______________.

A to get bigger

B to get cold

C to go into the air

D to turn to liquid

2. We wanted to know how hot it was outside, so we found out the temperature. Temperature is ______________.

A how fast the air is moving

B the amount of rain

C a tool for measuring

D how warm the air is

Name ____________________

Date ____________________

Lesson Quick Study

Lesson 1 - How Does Weather Change?

1. Inquiry Skill Practice–Observe

How did the weather change from morning to afternoon? What changes do you observe?

morning

afternoon

____

____

____

2. Use Vocabulary

Write the term that best completes each sentence.

seasons	**weather pattern**

A weather change that repeats is a ______________.

There are four ______________ in a year.

Name ______________________________

3. Focus Skill **Sequence**

Use this chart with the Reading Review.

spring The air gets warmer, and it is rainy.	→	Ⓐ __________ It is the warmest time of the year.	→	**fall** The air gets Ⓑ __________.	→	Ⓒ __________ It is the Ⓓ __________ time of the year.

4. **Critical Thinking and Problem Solving**

Suppose a friend sent you these pictures but did not tell you when they were taken. Use the clues in each picture to identify the season, and write it in the blank.

__________ __________ __________ __________

Name ______________________

Date ______________________

Lesson
Quick Study

Lesson 2 - Why Do We Measure Weather?

1. Inquiry Skill Practice–Compare

Which thermometer shows it is a cold day, and which shows it is a hot day? Label them hot and cold. Tell how you know.

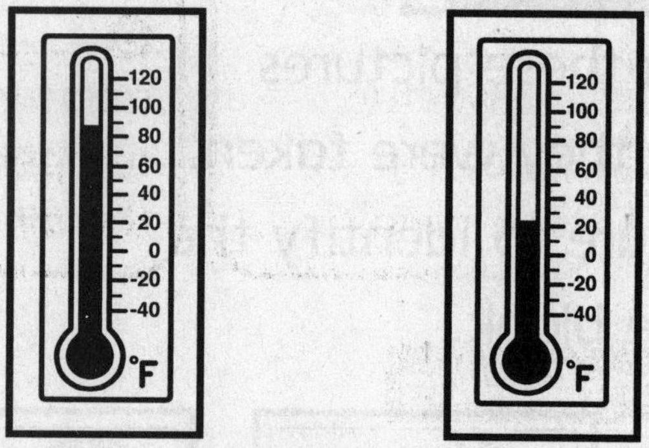

______________ ______________

____

____

2. Use Vocabulary

Write the word that best completes each sentence.

precipitation	**thermometer**

Rain and sleet are two kinds of ______________.

A ______________ measures how hot or cold the air is.

Name ______________________

3. Focus Skill **Main Idea and Details**

Use this chart with the Reading Review.

Main Idea and Details

Scientists use tools to measure weather.

- A (A) ______________ measures how much rain falls.
- An anemometer measures the speed of the (B) __________.
- A (C) ______________ shows the direction of the wind.
- A (D) ______________ measures the temperature of the air.

4. **Critical Thinking and Problem Solving**

Look at these pictures. Which one do you think shows the best wind for flying a kite? Tell why.

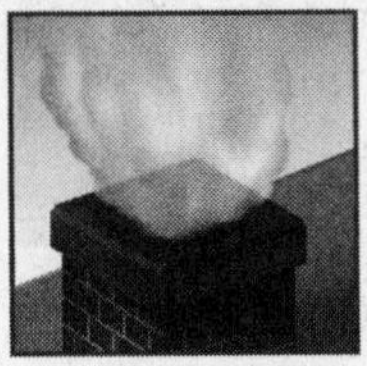

____

____

____

Name ______________________

Date ______________________

Lesson Quick Study

Lesson 3 - What Is the Water Cycle?

1. Inquiry Skill Practice–Infer

What happened to the puddle of water? Infer what caused the change.

____

____

____

2. Use Vocabulary

Match each word to the words that tell about it.

water cycle •	• Heat changes water to a gas.
drought •	• Water vapor cools and changes into water droplets.
condenses •	• This is all the movements of water between Earth's surface and the air.
evaporates •	• This is a long period with little or no rain.

Name ______________________________

3. Focus Skill **Cause and Effect**

Use this chart with the Reading Review.

cause	effect
The sun's heat warms water. →	Water Ⓐ ______________.
Water vapor meets cool air. →	Water vapor Ⓑ ______________.
Too little rain falls. →	There is a Ⓒ ______________.
Too much rain falls. →	There is a Ⓓ ______________.

4. **Critical Thinking and Problem Solving**

Suppose you are playing outside. You start to see dark clouds. You hear thunder. What do you think will happen? What should you do?

____

____

____

VOCABULARY GAMES

Contents

Vocabulary Games

You can use the word cards on pages RS 53–RS 80 to play these games.

Which One Does Not Fit?

You will need word cards

Grouping whole class, large group, or pairs

1. Choose three word cards. Two of them should fit a category. One of them should not fit that group.
2. Ask other players, "Which card does not fit?" Choose one player to tell why the word card does not fit the group.
3. Take turns putting word cards into groups. Guess which word card does not fit each group.

Word-O!

You will need 12 word cards for each player, 12 word cards for the caller

Grouping whole class, large group, or small group

1. Make three rows with your word cards. Each row should have three words in it. Place the words in any order face up. You should have three words that you do not use.

2. The caller will call one word at a time. If one of your words is called, turn it over.

3. When you have three words in a row turned over, you have won! Call out, "Word-O!"

ABC Ladder

You will need word cards

Grouping partners or small group

1. Place all of the word cards face up so that you can read each word.

2. Start making the ladder. Place the word card that comes first in ABC order.

3. Which word will come next in ABC order? Place it above the first word to form the ladder.

4. Take turns. Add word cards until all of them are placed in ABC order.

science tools	boulder
inquiry skills	investigate

boulder

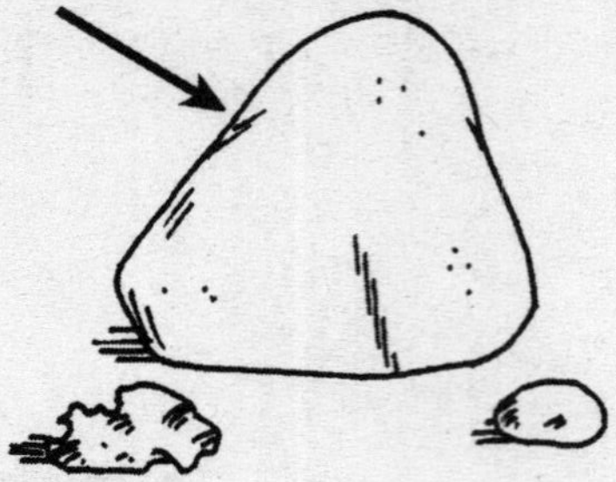

A **boulder** is a very large rock.

science tools

People use **science tools** to help find information they need. A thermometer is a **science tool**.

investigate

To **investigate** means to plan and do a test. Scientists **investigate** to answer a question.

inquiry skills

Inquiry skills are a set of skills people use to find out information.

erosion	**extinct**
dinosaur	**earthquake**

extinct

Dinosaurs are **extinct** because none of them lives on Earth anymore.

erosion

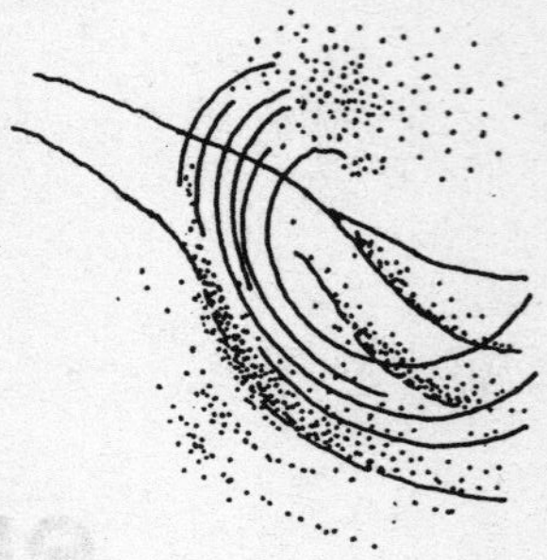

Erosion is the movement of sand and small rocks to a new place. Wind and water cause **erosion**.

earthquake

An **earthquake** is the shaking of Earth's surface that can cause land to rise or fall.

dinosaur

A **dinosaur** is an animal that lived on Earth millions of years ago. **Dinosaurs** have become extinct.

soil	**volcano**
fossil	**mineral**

volcano

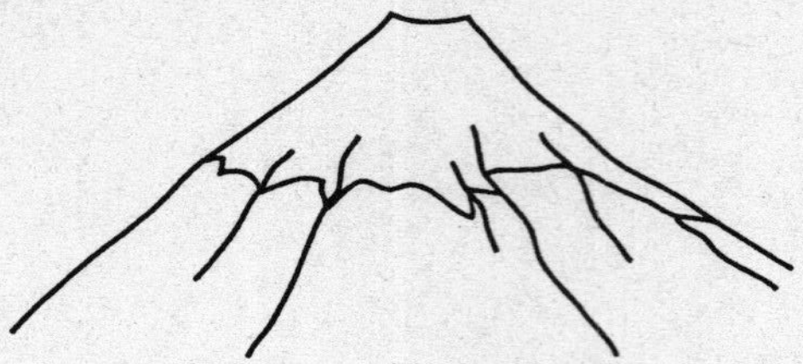

A **volcano** is a place where hot, melted rock, called lava, comes out of the ground onto Earth's surface.

soil

Soil is made up of bits of rocks mixed with matter that was once living. Farmers use **soil** to grow food.

mineral

A **mineral** is solid matter found in nature that was never living. Rocks are usually made of many different **minerals**.

fossil

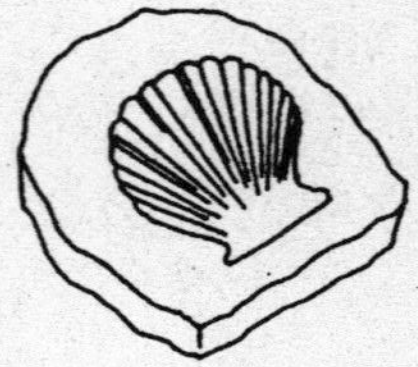

A **fossil** is what is left of a plant or an animal that lived long ago. A **fossil** can be a print in a rock or bones that have turned into rock.

natural resources

pollution

weathering

endangered

pollution

Pollution is waste that harms the air, water, or land. Fish in a river can be harmed by **pollution.**

natural resource

A **natural resource** is anything in nature, such as air, soil, and minerals, that people can use to meet their needs.

endangered

Endangered means in danger of dying out. People can help **endangered** plants and animals by protecting the places they live.

weathering

Weathering is a kind of change that happens when wind and water break down rock into smaller pieces.

resource	**reuse**
recycle	**reduce**

reuse

To **reuse** is to use a resource again.

resource

A **resource** is anything people can use to meet their needs. Trees and water are **resources.**

reduce

To **reduce** is to use less of a natural resource.

recycle

To **recycle** is to reuse the materials in old things to make new things. Plastic you **recycle** may become part of a new jacket.

friction

gravity

attract

force

gravity

Gravity is a force that pulls things toward the center of Earth. **Gravity** pulls a child down the slide.

friction

Friction is a force that slows down objects when they rub against each other. **Friction** also causes the objects to get warmer.

force

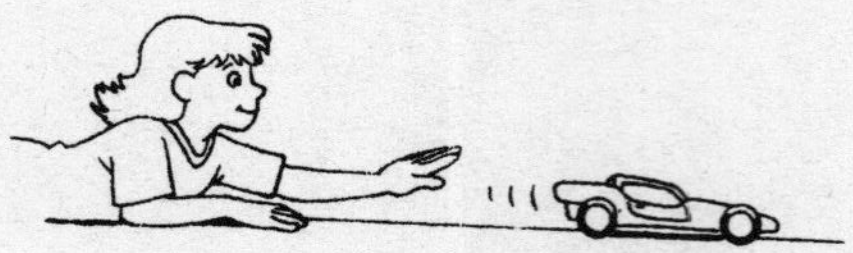

A **force** is a push or pull. Magnetism is one kind of **force**.

attract

To **attract** is to pull toward something. Opposite poles of two magnets **attract** each other.

pole

repel

magnet

motion

repel

To **repel** is to push away from something. Poles that are the same on two magnets **repel** each other.

pole

A **pole** is the end of a magnet. All magnets have a north-seeking pole and a south-seeking **pole**.

motion

Motion happens when something moves. When the girl jumps, she is in **motion**.

magnet

A **magnet** is an object that can pull things made of iron and steel. A **magnet** can push and pull other **magnets**.

nonliving

nutrients

speed

living

nutrients

Nutrients are substances plants and animals need to survive. Animals get **nutrients** from food. Plants get **nutrients** from the soil.

nonliving

Air, water, and the sun are **nonliving**, or not alive. A rock is a **nonliving** thing.

living

Plants and animals are alive, or **living**, because they need food, water, and oxygen. A puppy is a **living** thing.

speed

Speed is how fast something moves.

flowers

germinate

oxygen

shelter

germinate

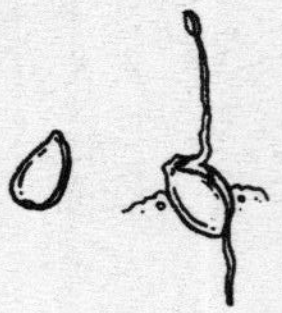

To **germinate** means to start to grow. A seed **germinates** when it gets water, warmth, and oxygen.

flowers

Flowers are plant parts that help a plant make new plants. Part of a **flower** makes seeds that grow into a new plant.

shelter

Shelter is a safe place to live. Bears may use a cave for **shelter**.

oxygen

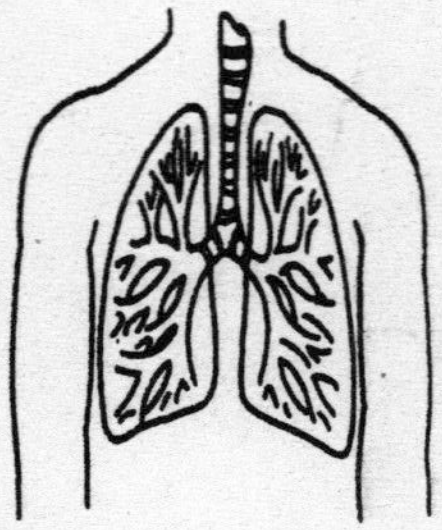

Oxygen is a gas in air and water. Your lungs take in **oxygen** when you breathe.

roots	**shrub**
leaves	**life cycle**

shrub

A **shrub** is a kind of plant that has many woody stems.

roots

Roots are the plant parts that take in water and nutrients. Most **roots** grow underground and help hold the plant in place.

life cycle

A **life cycle** is all the stages of a plant's or an animal's life.

leaves

Leaves are the plant parts that make food for the plant. **Leaves** use light, air, and water to make food.

condense

drought

stems

trunk

drought

A **drought** is a long time without rain. During a **drought**, the land may become dry, and plants may die.

condense

To **condense** is to change from water vapor, a gas, into liquid water. Water vapor **condenses** when heat is taken away.

trunk

A **trunk** is the one main stem of a tree.

stems

Stems are the parts of a plant that carry water and nutrients from the roots to the leaves.

season	**temperature**
evaporate	**precipitation**

temperature

Temperature is the measurement of how warm something is.

season

A **season** is a time of year that has a certain kind of weather. The four **seasons** are spring, summer, fall, and winter.

precipitation

Precipitation is water that falls from the sky. Rain, snow, sleet, and hail are kinds of **precipitation**.

evaporate

To **evaporate** is to change from liquid water into water vapor, a gas. Water **evaporates** when heat is added.

weather	**weather pattern**
thermometer	**water cycle**

weather pattern

A **weather pattern** is a change in the weather that repeats.

weather

Weather is what the air outside is like. The **weather** in spring is often rainy and warm.

water cycle

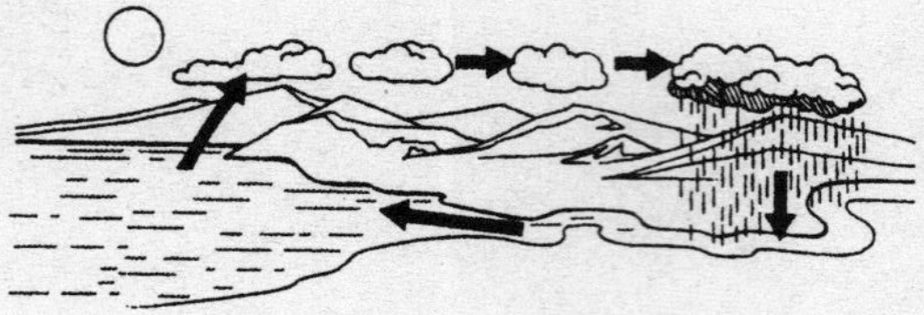

The **water cycle** is the movement of water from Earth's surface into the air and then back to Earth's surface.

thermometer

A **thermometer** is a tool used to measure temperature.

wind

wind

Wind is air that is moving. **Wind** can move sailboats.